AHRENS-FOX

FIRE APPARATUS PHOTO ARCHIVE

Walter M. P. McCall

Iconografix

Iconografix
PO Box 446
Hudson, Wisconsin 54016 USA

Library of Congress Control Number: 2008930628

ISBN-13: 978-1-58388-217-7
ISBN-10: 1-58388-217-0

08 09 10 11 12 13 6 5 4 3 2 1

Printed in China

Cover and book design by Dan Perry

Copyediting by Andy Lindberg

Cover photo-1927 Ahrens-Fox M-X-4 Piston Pumper #1314, ex-Bristol, Connecticut.
David R. White

DEDICATION

This book is respectfully dedicated to the memory of the late John F. Sytsma—America's No. 1 Ahrens-Fox aficionado, who through his photography and three books, shared his enthusiasm and vast knowledge of this iconic company and its distinctive products with all of us. Rest in peace, friend .The fire engines with the big silver ball up front must surely *protect your celestial domain. WMC*

BIBLIOGRAPHY

American Fire Engines Since 1900—By Walter M. P. McCall, Crestline Publishing, 1976

Ahrens-Fox: A Pictorial Tribute To A Great Name In Fire Apparatus—Edited by Robert Sams and John F. Sytsma, 1971

Ahrens-Fox Album—By John F. Sytsma, 1973

Ahrens-Fox Registration Numbers Of Motorized Fire Apparatus 1912-1957—By John F. Sytsma, 1974

Ahrens-Fox: The Rolls-Royce Of Fire Engines—By Ed Hass, 1982

F.D.N.Y. Ahrens-Fox Apparatus—By John F. Calderone, FAJ Publishing, 1991

Encyclopedia Of American Fire Engine Manufacturers—By Walter M. P. McCall, Iconografix, Inc., 2007

Photographs—Most of the photos in this book came from the substantial private collection of well-known fire apparatus historian Matthew Lee, to whom we are sincerely grateful for their use. An *MLC* (*Matt Lee Collection*) credit line identifies these photos. Others, including a number taken by the late John F. Sytsma, come from the author's own collection and are identified with a *WMC* or *JFS* credit line. Thanks, too, to Len Williams, Dan Jasina, Chuck Madderom, Jack Calderone, Shaun Ryan, Ken Buchanan, Frank DeGruchy, Martin Goffe, Jim Murray and Bob Stinson for their individual contributions to this project.

INTRODUCTION

Although this book's title is similar to those of the numerous others in Iconografix' *Photo Archive* series, it could just as well have been called *An Ahrens-Fox Primer.* Because that's what this book is—an illustrated overview of the many types of motor fire apparatus the Ahrens-Fox Fire Engine Company of Cincinnati, Ohio, produced during its fabled 45-year history.

This book in *no way* attempts to emulate the superb Ahrens-Fox pictorials published by the late John F. Sytsma in the 1970s, nor does it include anywhere near the wealth of minutiae on the company's products and evolution found in Ed Hass' comprehensive Ahrens-Fox history published in 1982. Self-published in small quantities and out of print for decades, John Sytsma's coveted Ahrens-Fox books are exceedingly hard to find today, and more than a quarter of a century has passed since Ed Hass published his fine history.

Although smaller than its better-known rivals in the U.S. motor fire apparatus industry—American-LaFrance, Seagrave and Mack—Ahrens-Fox nonetheless occupies a special place in the hearts and minds of fire apparatus buffs and historians everywhere. Lineal successor to America's first successful steam fire engine builder, Ahrens-Fox is best known for its highly distinctive piston-type pumping engines—the ones with the big silver ball atop their fully-exposed, front-mounted pumps.

Incorporated in 1910, the *Ahrens-Fox Fire Engine Company* was the successor to the *Ahrens Fire Engine Co.* formed in Cincinnati in 1905. One of the oldest and most respected makers of steam fire engines in the U.S., *Ahrens Mfg.* was a descendent of *A.B. and E. Latta*—the Cincinnati company that built the first successful steam fire engine in America in 1852.

John P. Ahrens and Charles H. Fox marched to their own drummer when they designed their first automobile pumping engine in 1910-11. Industry leader American-LaFrance chose a midship-mounted *rotary gear* pump, rival Seagrave a rear-mounted *centrifugal.* With its front-mounted *piston pump*—the same type long used on the company's highly regarded steam fire engines—the Ahrens-Fox *Continental* looked like no other.

"We delight in the fact that this engine differs from all others," the company boasted in a 1913 sales catalog. *"Boldly placed at the front, staunchly fortified by a fender of steel, the position of the pump is unique but—more to the point—logical."*

Alas, Ahrens-Fox's prototype motor pumper was the company's only notable failure. The pumper failed to meet its rated pumping capacity at the 1911 International Association of Fire Engineers' Convention in Milwaukee, Wisconsin. That pump was replaced with a much improved new one topped with two big side-by-side air domes. The first *Continental "A"* piston pumper was delivered to Rockford, Illinois, in early 1912. Forty-five more were sold over the next three years.

The Ahrens-Fox piston pumper redeemed itself in convincing fashion at the landmark 1913 IAFE motor pumping competition in New York where the company's *Continental Model "A"* was the *only* pumper to post a perfect score!

In 1914, Ahrens-Fox thoroughly redesigned its piston pumper. The most visible change was the replacement of the twin air domes previously used with a single *spherical* air chamber atop the pump. That big silver ball would become the Ahrens-Fox's most distinctive identifying feature for the next 38 years. The purpose of the air chamber (and the tall, cylindrical vacuum chamber in front of it) was to dampen the pulsation inherent in the vertical, double-acting reciprocating piston-type pump.

Over succeeding decades, a continuing series of piston pumpers bearing sequential alphanumeric model designations—*Models K, L. M, N, P* etc. came down the line. Pumping capacity ranged from 500 to 1,300 gallons-per-minute. The 1920s marked Ahrens-Fox's zenith, with city service ladder trucks, aerial trucks and quadruple combinations as well as rotary-gear pumpers joining the *Ahrens-Fox Standard Line.*

Ahrens-Fox barely survived the Depression. Deeply in debt for chassis purchased from the LeBlond-Schacht Truck Co., Ahrens-Fox was forced into a 1936 merger with Schacht for survival. The company moved out of its original factory on Colerain Avenue and into the LeBlond-Schacht plant. It was mostly downhill from there. Also during the 1930s, competitive pressure required Ahrens-Fox to offer centrifugal pumpers as well as its piston and rotary-gear types.

Because of its parent company's commitment to machine tool production, Ahrens-Fox suspended fire apparatus production during the Second World War. By the time peacetime production resumed in 1946, the piston pump was an anachronism—a relic from the past. Sales plummeted to a just a few a year. The last one was delivered in 1952. Despite the introduction of a handsome new custom conventional, Ahrens-Fox fell further behind its thriving competitors. Following two more reorganizations, Ahrens-Fox fire engine production was farmed out to bus builder C.D. Beck in Sidney, Ohio, in 1953. The company's last hurrah was the surprise introduction of a modern *cab-forward* pumper in early 1956. Just as production was ramping up, Mack Trucks acquired C.D. Beck, and the excellent cab-forward design was appropriated as Mack's highly successful C-Series. The last few fire engines to bear the Ahrens-Fox name were built in 1957.

Ahrens-Fox innovations included the "booster" water tank system (1913) which ultimately replaced the chemical system; the double-bank ladder bed (1923), and the Dahill pneumatic aerial hoist. Over its 45 years of existence Ahrens-Fox produced more than 1,500 pumpers, ladder trucks and other types of apparatus. Although half a century has passed since the last one was delivered, no other fire apparatus manufacturer in America has acquired the cult-like devotion accorded its famed piston pumpers.

There are two notable postscripts to the Ahrens-Fox story. In 1961, longtime Ahrens-Fox employee Richard C. (Curt) Nepper acquired the Ahrens-Fox name, records and remaining parts inventory—which he utilized to keep countless Ahrens-Fox fire engines running and pumping. In 1970 Nepper's *Ahrens-Fox Company* was commissioned to build a *new* Ahrens-Fox pumper on tilt-cab Ford chassis for Southern Campbell, Kentucky. Because of compromises insisted on by the purchaser, Mr. Nepper deemed the apparatus unworthy of the Ahrens-Fox name when it was delivered seven years later.

In 1993, Ahrens-Fox devotee Kenneth Menke of Webster Grove, Missouri, purchased the Ahrens-Fox Company from the ailing Mr. Nepper, who passed away shortly thereafter. To provide public access to this storehouse of information and parts, Menke formed the *Ahrens-Fox Owners' Association.* In 2002, Mr. Menke leased this iconic name to HME Fire Apparatus of Wyoming, Michigan, for exclusive use on HME's premium custom fire apparatus

Nearly 100 years after the Ahrens-Fox name was minted in Cincinnati, this revered marque still resonates with the credo of its famed founders—*"Nothing too good for the fire service, only the best will do."*

AHRENS-FOX HISTORICAL HIGHLIGHTS

1905	*Ahrens Fire Engine Co.* begins operation in new factory in Cincinnati, Ohio
1910	Name changed to the *Ahrens-Fox Fire Engine Company*
1911	Prototype motor-driven piston pumper fails IAFE test in Milwaukee
1912	First double-dome *Continental Model "A"* delivered to Rockford, Illinois
1913	Ahrens-Fox *Booster Auxiliary* system introduced, on Republic truck chassis
1913	Model "A" *Continental* posts only perfect score at IAFE pump test, New York
1914	Ahrens-Fox begins making its own four- and six-cylinder motors
1915	First spherical dome piston pumper delivered to Beaumont, Texas
1916	Last Ahrens-Fox steamer built on White truck chassis for Lansford, Pennsylvania
1919	New shaft-drive, gable-hood Ahrens-Fox introduced
1923	Aerial and city service ladder trucks join Ahrens-Fox line, Dahill air hoist
1925	First Ahrens-Fox *Rotarystyle* rotary-gear pumper built
1928	*Skirmisher* quad and *Satellite* rotary gear pumpers added to line
1930	Schacht-based *Model "V" Junior Line* introduced
1936	Merger with LeBlond-Schacht, Ahrens-Fox moves into L/S Evans Ave. plant.
1937	Curt Nepper-designed *Model HT* 1,000-gpm piston pumper introduced
1939	Ahrens-Fox again uses Schacht chassis for rotary, centrifugal pumpers
	Production transferred to LeBlond plant in Norwood, Ohio
1940	Ahrens-Fox suspends fire apparatus production during Second World War
1946	Ahrens-Fox resumes motor fire apparatus production
1947	New postwar conventional Ahrens-Fox design introduced
1950	Assets sold to Cleveland Automatic Machine Co., Norwood, Ohio
1951	Ahrens-Fox Division sold to General GMC Truck Sales, Cincinnati. Ohio
1952	Last Ahrens-Fox piston pumper delivered to Tarrytown, New York
1953	General Truck subcontracts Ahrens-Fox production to C.D. Beck, Sidney, Ohio
1955	New Beck-built cab-forward *FCB* fire apparatus announced
1956	Mack Trucks acquires C.D. Beck bus and fire apparatus operations
1957	Last Ahrens-Fox fire engines built
1961	Richard C. (Curt) Nepper acquires Ahrens-Fox name, assets, and parts inventory
1977	Curt Nepper delivers Ford C-Series pumper to Southern Campbell, Kentucky.
1993	Kenneth Menke, Webster Grove, Mo. purchases Ahrens-Fox Company from Curt Nepper
2002	Ahrens-Fox name revived on premium series HME custom fire apparatus.

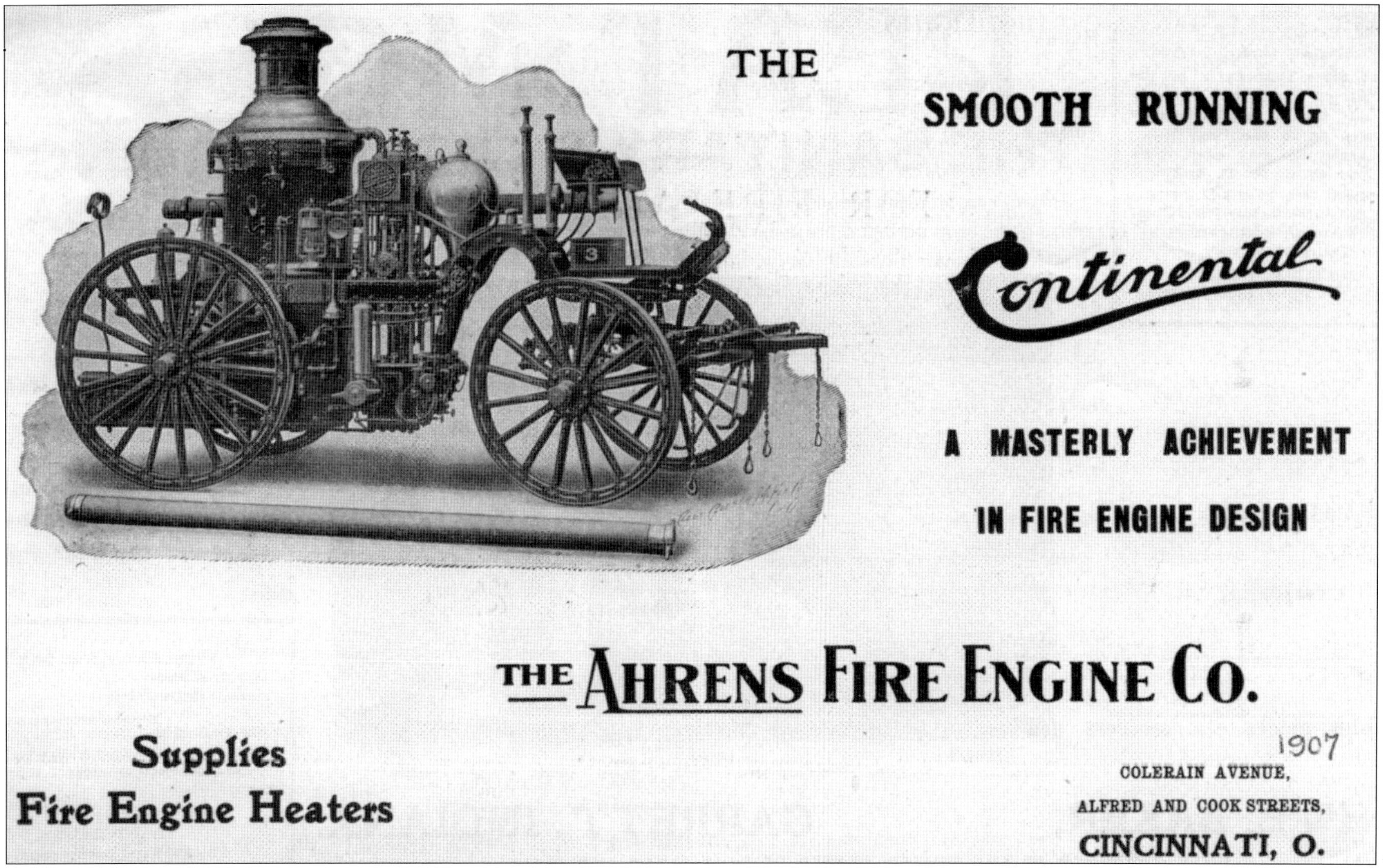

1907 Ahrens *"Continental"* Steamer—The principal product of the newly formed (1905) *Ahrens Fire Engine Company* was the *"Continental"* horse-drawn steam fire engine. Designed to compete with such state-of-the-art steamers as American-LaFrance's *Metropolitan,* the *Nott Universal* and *Amoskeag,* the *Ahrens Continental* was built in six sizes—from the 500 gallon-per-minute *Fourth Size* on up to the mighty 1,300 gpm *Double Extra First Size.* Ahrens and Ahrens-Fox built more than 100 *Continental* steamers before the last one was delivered in 1916. The highly regarded *"Continental"* name was wisely carried over onto the company's new motor-driven fire apparatus in 1912. *WMC*

1911 Ahrens-Fox Piston Pumper Prototype—The Ahrens-Fox Fire Engine Company began design work on its first automobile-pumping engine not long after the company adopted its new hyphenated name in 1910. By 1911 the design had been finalized and a prototype constructed. Bearing factory serial number *500,* the first motorized Ahrens-Fox pumper was equipped with a front-mounted 450 gallon-per-minute piston pump—which failed to meet its intended capacity. This pump, with a single onion-shaped air chamber, was replaced with an improved pump with two big air domes. Assigned new serial number *501*, the first Ahrens-Fox *Continental* piston pumper was delivered to Rockford, Illinois, in early 1912. *MLC*

Double-Dome Pump—Introduced in early 1912, the redesigned Ahrens-Fox *Continental Model "A"* had a front-mounted 750 gallon-per-minute, two-piston pump surmounted by two large dome-shaped air chambers. Wheelbase was 148 inches. Ahrens-Fox delivered 46 piston pumpers of this style between 1912 and 1915. These massive, bulky-looking fire engines were powered by an 80-horsepower six-cylinder Herschell-Spillman motor with a 5¾" cylinder bore and 6" piston stroke. *MLC*

Above—Detroit, Michigan, was Ahrens-Fox's first volume purchaser of motor fire apparatus. The Detroit Fire Department received nine Ahrens-Foxes in 1912 and 1913—four *Continental Model "A"* piston pumpers, four *Model "F"* tractors to motorize formerly horse-drawn ladder trucks, and the only *Model "C" Combination Hose Car* ever built. This is the first of the four 700 gallon-per-minute piston pumpers. **Below—**After years of service the *Model "C"* high-pressure hose wagon was converted into a ladder truck tractor. Note the engine-driven power generator mounted in front of the radiator. *MLC*

Three Models—Ahrens-Fox *Continental* piston pumpers were built in three versions. *Model "A"* was a dual-combination pump and hose car, *Model AC* a triple combination pump, hose and chemical car, and *Model "B"* a straight pumping engine with crew seats in what would normally have been the hose bed. **Above**—1912 *Continental "B,"* one of two built for Washington, D.C. Note flared front fenders and chain-drive sprockets *behind* the rear wheels. **Below**—1912 *Continental Model "A"* #510 was still in service in Edgewater, New Jersey, when this photo was taken in the late 1930s. *MLC*

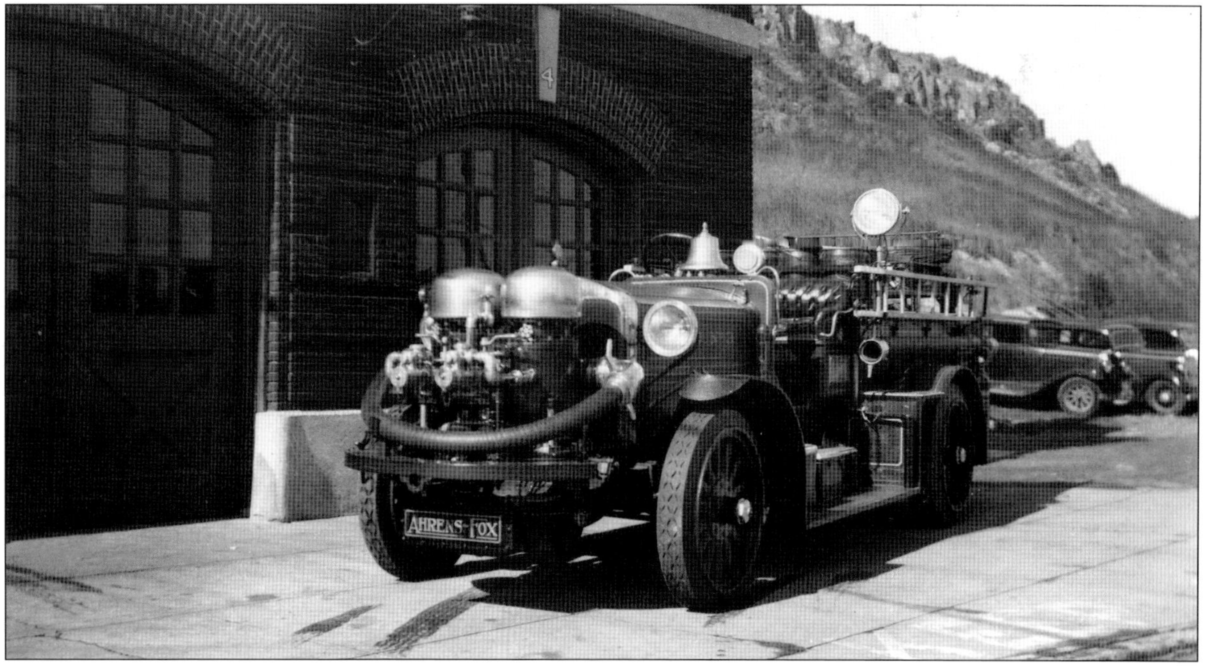

Clean Sweep—The Ahrens-Fox piston pumper proved itself in convincing fashion at the historic 12-hour pumping demonstration held during the 1913 International Association of Fire Engineers' Convention in New York City. The Ahrens-Fox *Model "A" Continental* was the *only* pumper to post a *perfect score.* Following the landmark IAFE convention, #519 was delivered to New Bedford, Massachusetts. This photo, taken during the pumping test, shows the twin air domes, massive steel bumper and prominent *Ahrens-Fox* nameplate. Note the side-mounted suction intakes and the three throbbing 2½" hose lines coming off the front of the pump. *MLC*

Above—Still rolling on hard rubber tires, 1914 Ahrens-Fox *Model "A" Continental* #539 was photographed near the end of its long service life in Sacramento, California. The last double-dome Ahrens-Fox *Continental* piston pumper was delivered to Danvers, Massachusetts, in mid-1915. **Below**—New Bedford's 1913 *Model "A" Continental* at the FDNY Repair Shops prior to the historic 12-hour pumping competition. *MLC*

1913 Model "D" Booster Car—One of Ahrens-Fox's most significant contributions to the fire service was the *Booster Auxiliary* system introduced in 1913. A small front-mounted pump was connected to a 60-gallon water tank that replaced the usual soda-and-acid chemical system. The hard rubber *booster hose* was carried on a reel under the driver's seat. The Cincinnati Fire Department purchased ten of these speedy little *Model "D" Squad Cars* built by Ahrens-Fox on light-duty chassis purchased from the Republic Motor Car Co. of Hamilton, Ohio. By the 1930s, the booster system had all but replaced the chemical system for fighting small fires. Twenty-two Ahrens-Fox/Republics were sold between 1913 and 1915. *MLC*

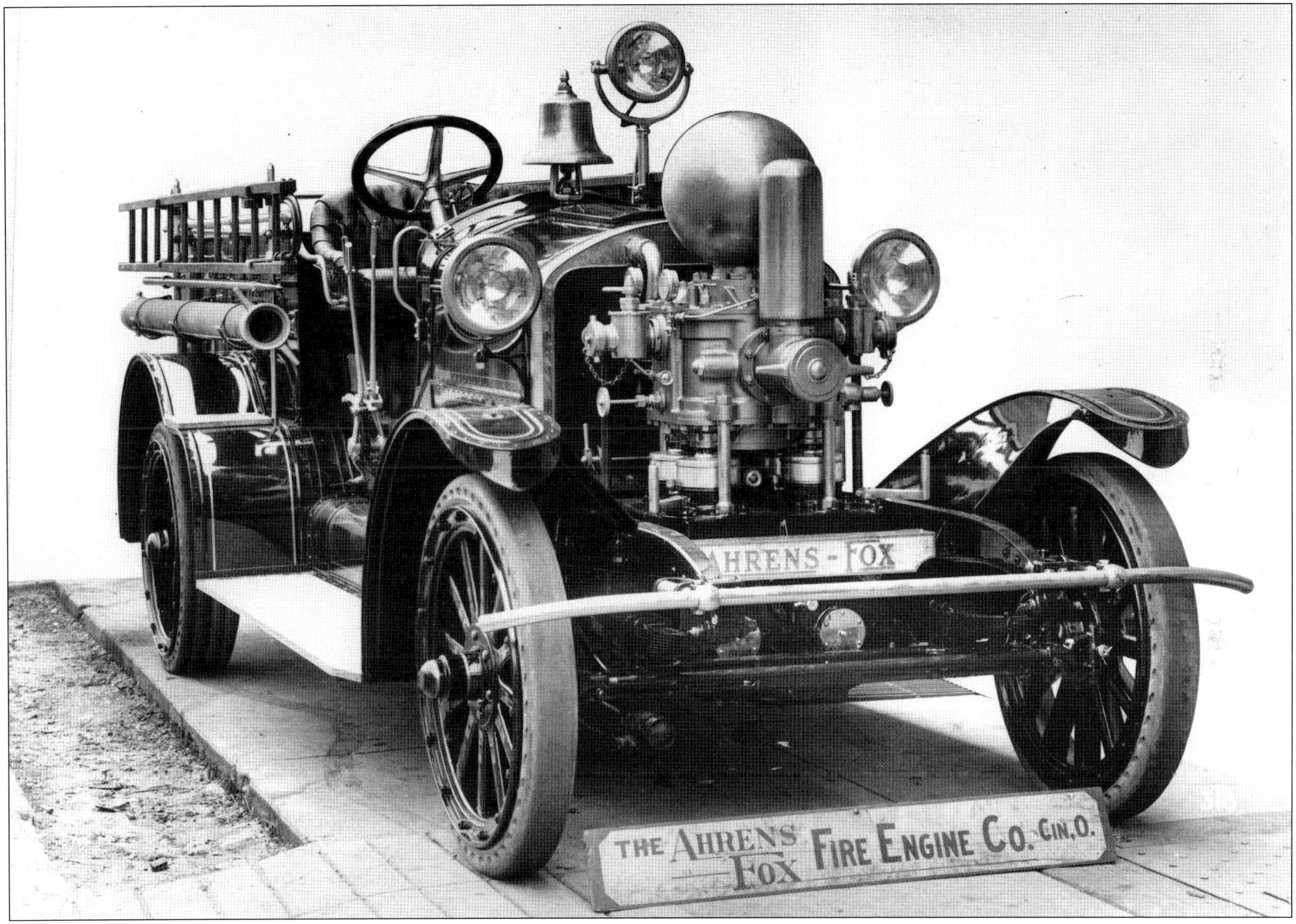

1914 Model K-2 Piston Pumper—Ahrens-Fox began making its own four- and six-cylinder motors in 1914. At this time the company also thoroughly redesigned its front-mounted piston pump. The most visible change was the replacement of the two big air chambers used on the earlier *Continental* piston pumpers with a single *spherical* air chamber atop the pump. The big, silver ball and the tall, cylindrical surge tank in front of it would become unmistakable hallmarks of Ahrens-Fox fire engines for the next four decades. Built in 1914 and exhibited at conventions that year, the first spherical-dome Ahrens-Fox piston pumper, Model K-2 Reg. #601, was sold to Beaumont, Texas, in 1915. Note also the single front suction intake. *MLC*

1915 Model K-14 Combination—The increasingly successful Ahrens-Fox Fire Engine Company introduced three new products in 1915—all with sequential alphanumeric model designations. The *Model "K"* was a line of four-cylinder piston pumpers, small combination hose and chemical cars and four-wheel ladder truck tractors. *Model "L"* was a 1,100-gpm six-piston pumper and *"Model M"* a four-cylinder 750-gpm pumper. Covington, Kentucky, received two 1915 *Model K-14 Combination Hose & Chemical Cars* (#603 and 604), one of which is shown here. Note the rear-mounted chemical hose reel and stubby wheelbase of this compact fire apparatus. *MLC*

1915 Model K-11 Combination—Chicago Heights, Illinois, got the first Ahrens-Fox *Model K-11 Combination Hose and Booster Car, #602,* in early 1915. Powered by the company's new 80-horsepower four-cylinder engine, the K-11 had a front-mounted 250-gpm rotary gear pump and 60-gallon booster tank. Note the dual booster hose reels, the electric siren on the dash and the home town-boosting *Made In Cincinnati* banner tied to the short wooden ladder. Attired in their rubber turnout coats and high-front helmets, members of the Chicago Heights Fire Department pose with their just-delivered K-11 during its acceptance test. *MLC*

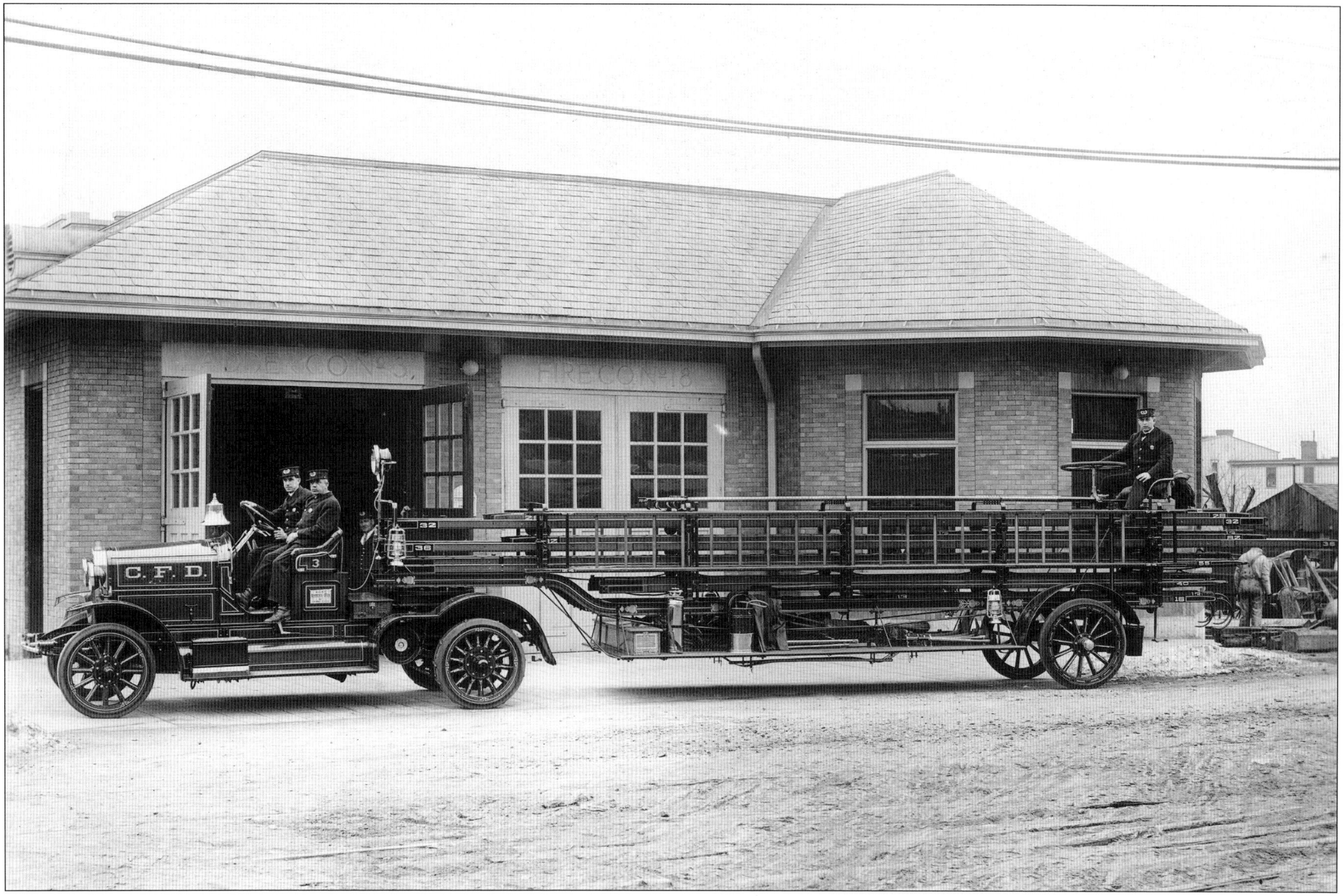

1916 Model K-17 Tractor—Not surprisingly, Ahrens-Fox's home city of Cincinnati, Ohio, was a good customer. In the `teens and twenties the Cincinnati Fire Department replaced most of its horse-drawn fire apparatus with a large fleet of Ahrens-Fox pumpers and ladder truck tractors. This is one of a dozen four-cylinder *Model K-17* tractors purchased to haul formerly horse-drawn city service hook and ladder trucks and aerials. Reg. #661 is attached to Ladder Company #3's city service ladder truck. Note the swivel spotlight at the front of the single-bank trailer, and the tillerman's lofty perch. *MLC*

1916 Model "L" Piston Pumper—First in a continuing series of high-capacity piston pumpers built by Ahrens-Fox was the 1,100 gallon-per-minute *Model "L"* introduced in 1915. Powered by a 130-horsepower six-cylinder engine of $5^{7}/_{8}$" bore and 7" stroke coupled to a six-piston pump, the mighty "L" was Cincinnati's answer to rival American-LaFrance's rotary gear *Type 15.* Pottstown, Pennsylvania, received this *Model L4, #705,* in 1916. *Robrecht Negatives*

First Ahrens-Fox Aerial Ladder Truck—Charles H. Fox was a good friend of Edward F. Dahill, Chief of the New Bedford, Massachusetts, Fire Department and inventor of the *Dahill Pneumatic Aerial Ladder Hoist.* Ahrens-Fox constructed this 75-foot straight-frame aerial with Dahill air hoist and four-wheel electric drive in 1916. Reg. #429 was built on an all-wheel-drive chassis purchased from the Couple Gear Freight Wheel Company of Grand Rapids, Michigan. A generator in each wheel provided smooth, eerily quiet—but slow—propulsion. Two years later Ahrens-Fox built a Couple Gear electric-drive piston pumper for Gloucester, Massachusetts. That application proved far less successful however and the pumper was soon converted to conventional gasoline power. New Bedford's aerial remained in service into the 1940s. Ahrens-Fox did not regularly start building aerials until 1923. *WMC*

1916 Ahrens-Fox MK2 Piston Pumper—This catalog illustration of a typical Ahrens-Fox piston pumper of the First World War era shows the distinctive configuration Ahrens-Fox successfully utilized for four decades. The driver of this dual-combination Pump and Hose Car sits virtually amidships. The placement of the four-piston pump at the front of the apparatus, ahead of the radiator, enabled the driver to precisely "spot" the pumper at a hydrant, or drive right up to the edge of the water source for drafting. Note the louverless two-section hood side panels. Introduced in 1915, the 800 gallon-per-minute *Model MK* was powered by a 90-horsepower six-cylinder engine of 5½" bore and 6½" stroke coupled to a *Twin Double Multiplex* four-cylinder pump. The *MK2* carried no chemical or booster equipment. *MLC*

1917 K2 Piston Pumper—In 1915, Ahrens-Fox replaced its bulky first-generation *Continental "A"* piston pumper with a thoroughly redesigned, somewhat smaller "K" Series powered by an 80-horsepower four-cylinder engine. The pleasingly styled "K" sported a single spherical air chamber over the four-cylinder 750 gallon-per-minute pump instead of the two side-by-side air domes used on the more than 40 *Continentals* built between 1912 and 1915. This is an in-service photo of a 1917 Ahrens-Fox Model K2 delivered to New Bedford, Massachusetts. Note the Baker cellar pipe carried over the rear fender and the coil of rope dangling from the dash. Reg. #642's original hard rubber tires have been replaced with pneumatics. *MLC*

Cincinnati Stablemates—During the teens and 1920s, Cincinnati motorized its fire department with a huge fleet of Ahrens-Fox piston pumpers and ladder truck tractors. Most Cincinnati firehouses of this period housed nothing but Ahrens-Foxes. Taken in the early 1920s, this photograph is typical. On the left is 1917 *Model K-17* tractor #668 attached to Water Tower #2's 1896 Hale, and on the right Engine 21's 1917 *Model M4* piston pumper #832. *MLC*

1918 Model M-11-14 Special Combination Hose Car—Ahrens-Fox sometimes photographed completed apparatus before the machine went into the paint shop. Here's a good example. This *Model M-11-14 Special Combination Hose and Chemical Car,* Reg. #853, was built for Manchester, New Hampshire. Note the small front-mounted rotary gear pump and the flared *Springfield* hose body. This distinctive style of hose body was fitted with two parallel crew seats above the hose load. Note also the cover over the drive chains. *MLC*

Ahrens-Fox In Uniform—The United States Government purchased 15 Ahrens-Fox piston pumpers in 1918. Painted standard military olive drab, the *Model IK4Triple Combinations* with 750 gallon-per-minute pumps and booster equipment were posted to U.S. Army and Navy installations all over the country. This one, photographed in September 1918, was assigned to the U.S. Ordnance Department. Note the hand-cranked siren on the dash and the booster hose basket over the hose bed. Powered by an 80-horsepower Ahrens-Fox four-cylinder engine, these pumpers also had four-cylinder *Twin Double* pumps. *MLC*

1918 Model K-34 Rotary Gear Pumper—In addition to its bread-and-butter piston pumpers, Ahrens-Fox turned out some interesting oddities from time to time. This is one of them. In 1918, Bellevue, Kentucky, received this one-off *Model K-34 Pump & Hose Car, #690,* which was equipped with a small front-mounted *rotary gear* pump. This truly unique little Ahrens-Fox is owned today by fire apparatus collector Chip Lytle of Kettering, Ohio. *WMC*

1919 Model "P" Piston Pumper—Successor to the earlier *Model "L,"* Ahrens-Fox's new heavy-hitter was the *Model "P"* introduced in 1919. Powered by the company's largest six-cylinder engine—a 1,140 cubic inch, 130-horsepower motor which drove a front-mounted *Twin Triple* six-piston pump—the Model P put out an impressive 1,400 gallons-per-minute—even more than the Double Extra First-Size steam fire engine it succeeded. Just 23 Model "P"s were built over the decade between 1919 and 1929. This is one of a pair of IP2's (#721 and #722) delivered to Louisville, Kentucky, in 1919. *MLC*

1920 Model JM3 Piston Pumper—With the introduction of the 750 gallon-per-minute, six-cylinder Model "J" in 1919, the Ahrens-Fox piston pumper underwent another major appearance change. The rounded hood and radiator used since 1915 were gradually supplanted by a handsome new *gabled* hood with louvered side ventilation panels. But the most significant advance on this new model was the change from chain to *shaft drive.* Even after the introduction of this new front-end ensemble, the chain-drive, round hood style Ahrens-Fox continued in production until 1923. The classic gable hood era lasted 12 years—from 1919 to 1931. This is one of three Model JM3 pumpers delivered to the Detroit Fire Department in 1920. Note the unusual location of the chemical tank under the driver's seat. *MLC*

The "Fighting Face" Of The Ahrens-Fox Piston Pumper—Immortalized in a series of Ahrens-Fox ads published in fire service trade journals in the 1920s, this was the ferocious "fighting face" of the famed Ahrens-Fox piston pumper. The "model" for this photograph was a 1921 *Model JM4* 750-gpm piston pumper, #990, built for the local Cincinnati Fire Department. Note the single-piece spherical air chamber. A new two-piece "ball" with circumferential center band was introduced in 1923. *WMC*

Ahrens-Fox's Model "T"—Just like Henry Ford, Ahrens-Fox also had a *Model "T."* But unlike Ford's famed flivver, Ahrens-Fox's Model "T" was less than a resounding sales success with only two T-3 pumpers and one hose truck built. The four-cylinder Model "T" had a front-mounted 500 gallon-per-minute *inline* five-piston pump—without the usual air chamber. The other T-3 pumper was lost at sea on its delivery voyage to Soerobaya, Java, in the Dutch East Indies. Batesville, Indiana, still owns this 1921 *Model T-3* (#1002), which had been retrofitted with a windshield when the late Jim Burner photographed it many years ago. That's a siren atop the pump. *MLC*

Fours And Sixes—Depending on pump size, Ahrens-Fox piston pumpers were available with four- or six-cylinder engines. **Above:** 1922 four-cylinder 750-gpm *Model KS4* for Sugar Notch, Pennsylvania. **Below:** 1922 six-cylinder 1,000-gpm *Model NS3 # 927* on hard rubber tires for Newark, New Jersey. *MLC*

1922 Model J-13 High Pressure Hose Car—Ahrens-Fox designed and built two of these special high pressure hose cars for the Detroit Fire Department. Fitted with two huge turret nozzles and extra large hose beds for three-inch hose, these firefighting battlewagons were in active service for nearly 50 years. When the D.F.D. closed its high pressure pumping station in the 1960s, the Ahrens-Fox high pressure hose wagons were redesignated *Boat Tenders 2 and 3,* providing land support for the fireboat *John Kendall.* Retired in the early 1970s they're both still on the road today owned by well-known collectors Mike Adams of Burt, Michigan, and former Detroiter Len Williams of Sunnyvale, California. *WMC*

Ladder Truck Tractor—The purchaser and serial number of this Ahrens-Fox four-wheel tractor of early 1920s vintage are unknown. It's somewhat unique for a gable-hood Ahrens-Fox inasmuch as it has chain drive instead of the shaft drive seen on most gable hood Foxes. Note the *Sewell Solid Cushion* tires and the "fifth wheel" for attachment of an aerial ladder or city service hook and ladder truck trailer. The hood is lettered *N.F.D. No. 1.* The *Ahrens-Fox* nameplate in the foreground is the same one placed between the frame rails on the company's piston pumpers.

1923 Model IK4 Piston Pumper—Front wheels tucked in to the curb, Harvey, Illinois' 1923 Ahrens-Fox Model IK4 is seen here hooked up to a hydrant not long after it was placed in service. This would be the final year for the round hood, chain-drive Ahrens-Fox. The two 2½" hose lines from the hydrant are hooked into reducers in #1136's Y-shaped suction intake. Note the transverse-mounted booster tank behind the seat. Nicely restored, this four-cylinder 750-gpm triple combination pumper is today owned by the Rosemont (Illinois) Fire Department. *MLC*

A Fox Takes A Breather—The pump operator takes a break on the running board of his Ahrens-Fox piston pumper at an unknown location sometime in the mid-1920s. On many fire departments, standard operating procedure was to raise the hood side panels to maximize engine compartment ventilation while pumping, Note the six-inch hard suction hose attached to the pump. This is a gable hood, shaft-drive job. Note also the spotlight, oval fuel tank and large cylindrical booster tank above the hose box. *MLC*

Classic Factory Photos—The Ahrens-Fox Fire Engine Company used airbrushed photos like these, without backgrounds, in its sales literature and advertising. **Above:** 1925 *Model NS4* 1,000 gpm-piston pumper, #1643, Beverly, Massachusetts. **Below:** 1923 *Model JM4* 750-gpm pumper, #1218, one of three built for Long Branch, New Jersey. *MLC*

City Service Ladder Trucks— Ahrens-Fox charged into the ladder truck market in a big way in 1923, introducing a complete line of city service and aerial hook and ladder trucks. Here are two delivered that milestone year. **Above:** Single-bank six-cylinder *Model JS43* built for Dunkirk, New York. **Below:** Double-bank four-cylinder Model *KS43* for Braintree, Massachusetts. *MLC*

Aerial Ladder Trucks—Introduced in 1923, Ahrens-Fox tractor-drawn aerial ladder trucks were equipped with Dahill pneumatic hoists and two-section 75- or 85-foot wood aerial ladders sourced from the Combination Ladder Co. of Providence, Rhode Island. All had six-cylinder tractors. Here are two views of a 1923 *Model 85-6-1,* #2001, delivered to Nashua, New Hampshire, photographed on the park-like campus of the University of Cincinnati. *MLC*

Convention Exhibit—Ahrens-Fox could always be found front and center at major fire chiefs' conventions across the country. The company often displayed its piston pumpers, combinations and ladder trucks in a fenced-in enclosure in a prominent location on the convention show floor. The site of this particular show is not known. Note the modest claim, *Builders of the World's Best Fire Engine,* on the placard in front of this six-cylinder piston-type pumping engine. No argument here! *MLC*

Firehouse Background—The Ahrens-Fox company photographer often photographed new apparatus in front of this imposing Cincinnati firehouse, the quarters of Engine 43 and Ladder 13, just down the street from the factory on nearby Colerain Street. On-duty fire fighters often posed with rigs destined for distant cities, like this long 85-foot double-bank aerial—one of two shipped to Manchester, New Hampshire, in 1923. Note the angled lifting cylinders and dual compressed-air tanks on the trailer frame for the Dahill air hoist. *MLC*

Four-Cylinder Tractor—Ahrens-Fox offered its all-new city service hook and ladder trucks in two basic versions. Customers could specify a four-wheel straight-frame or six-wheel tillered tractor-trailer ladder truck powered by a four- or six-cylinder engine, with single- or double-bank ladder bed. Only six tractor-drawn service trucks were sold compared with deliveries of dozens of the conventional four-wheel, non-tillered type. Here's a close-up of the four-cylinder *Model KS17* tractor on Reg. #1145 delivered to Manchester, New Hampshire, in early 1924. *MLC*

1924 Model 75-6-1 Tractor-Drawn Aerial—Kokomo, Indiana received this 75-foot tractor-drawn Ahrens-Fox aerial, #2004, in 1924. The crew seems mighty proud of its new mount. Note the hard rubber tires on wood-spoked artillery wheels and the big Vesta drum headlights. This aerial, which was later converted to pneumatic tires, is owned today by Mike Adams of Burt, Michigan. *Mike Adams*

The Mighty PS4—Successor to the earlier chain-drive *Model "L,"* Ahrens-Fox's most powerful piston pumper of the 1920s was the shaft-drive, gable-hood *Model PS4.* Powered by a 1,140 cubic inch, 130-horsepower Ahrens-Fox six-cylinder engine of 5⅞" bore and 7" stroke coupled to a 1,300 gallon-per-minute *Twin Triple* six-piston pump, the PS4 competed head-to-head with rival Seagrave's 1,300-gpm *Metropolite* centrifugal and American-LaFrance's 1,400-gpm rotary gear *Type 15.* Passaic, New Jersey, which purchased a large fleet of high-capacity Ahrens-Fox pumpers in the 1920s, received PS4 #729 in May 1923. *MLC*

SPAAMFAA'S "Big Bertha"—Syracuse, New York, was the original purchaser of this 1,300 gallon-per-minute *Model PS4* twin-triple piston pumper #728, delivered in 1923. After passing through several owners following its retirement in 1952, "Big Bertha" was acquired and restored by the Syracuse-based *Society for the Preservation and Appreciation of Antique Motor Fire Apparatus in America*—better known as *SPAAMFAA.* Painted the S.F.D.'s distinctive white and red with lots of golf-leaf striping and trim, #728 was recently restored from the frame up for the second time—just in time for the Society's 50th Anniversary celebration in Syracuse in 2008. *JFS*

1924 Model PS2—San Antonio, Texas, received two 1,300 gallon-per-minute *Model PS2's* (#730 and 731) in October 1924. Later used in the company's advertising, this remarkable photo shows one of these powerful piston pumpers during its acceptance test in downtown San Antonio. Six-inch soft suction hooked up to a large-main hydrant, and with wyes on all four discharge gates, the PS2 is supplying no fewer than *eight* 2½" hose lines—performance befitting the company's boast of making, *"The pump with the punch!" MLC*

Six-Cylinder Service Trucks—Here are two more Ahrens-Fox double-bank city service hook and ladder trucks, both built in 1924. **Above:** Six-wheel tillered tractor-drawn *Model JS17-44,* Reg. #1224 for Attleboro, Massachusetts. Note soda and acid chemical tank on front of trailer. **Below:** Four-wheel, straight-frame *Model NS44,* Reg. #1613 with chemical equipment for Duluth, Minnesota. The *NS44* was the company's most popular city service ladder truck. *MLC*

1925 Rotary Gear Pumper—Up until the mid-1920s, Ahrens-Fox built *only* piston-type pumpers. But many fire chiefs preferred the rotary gear and centrifugal pumps offered by rivals American-LaFrance and Seagrave. Aware of this growing competitive pressure, Ahrens-Fox developed its own *Rotarystyle* rotary gear pump. Although the first Ahrens-Fox midship-mounted rotary gear pumper, *Model G-J-5-4 #5001,* was delivered to Danvers, Massachusetts in 1925, the company did not begin volume production of rotary gear pumpers for another three years. *MLC*

1925 Salvage Patrol—In addition to its *Standard Line* of pumpers and ladder trucks, the Ahrens-Fox Fire Engine Co. designed and built some interesting special-purpose apparatus for its customers. Among them was this big six-cylinder *Model PS-16 Salvage Patrol Car, #735,* built in 1925 for the *Worcester Protective Department* in Worcester Massachusetts. The hood is lettered *Fire Patrol.* Note the compartmented rear body, equipment lockers on the running board, canvas windbreaker fitted around the bell, and the spare tire lashed to the side of the driver's seat. *MLC*

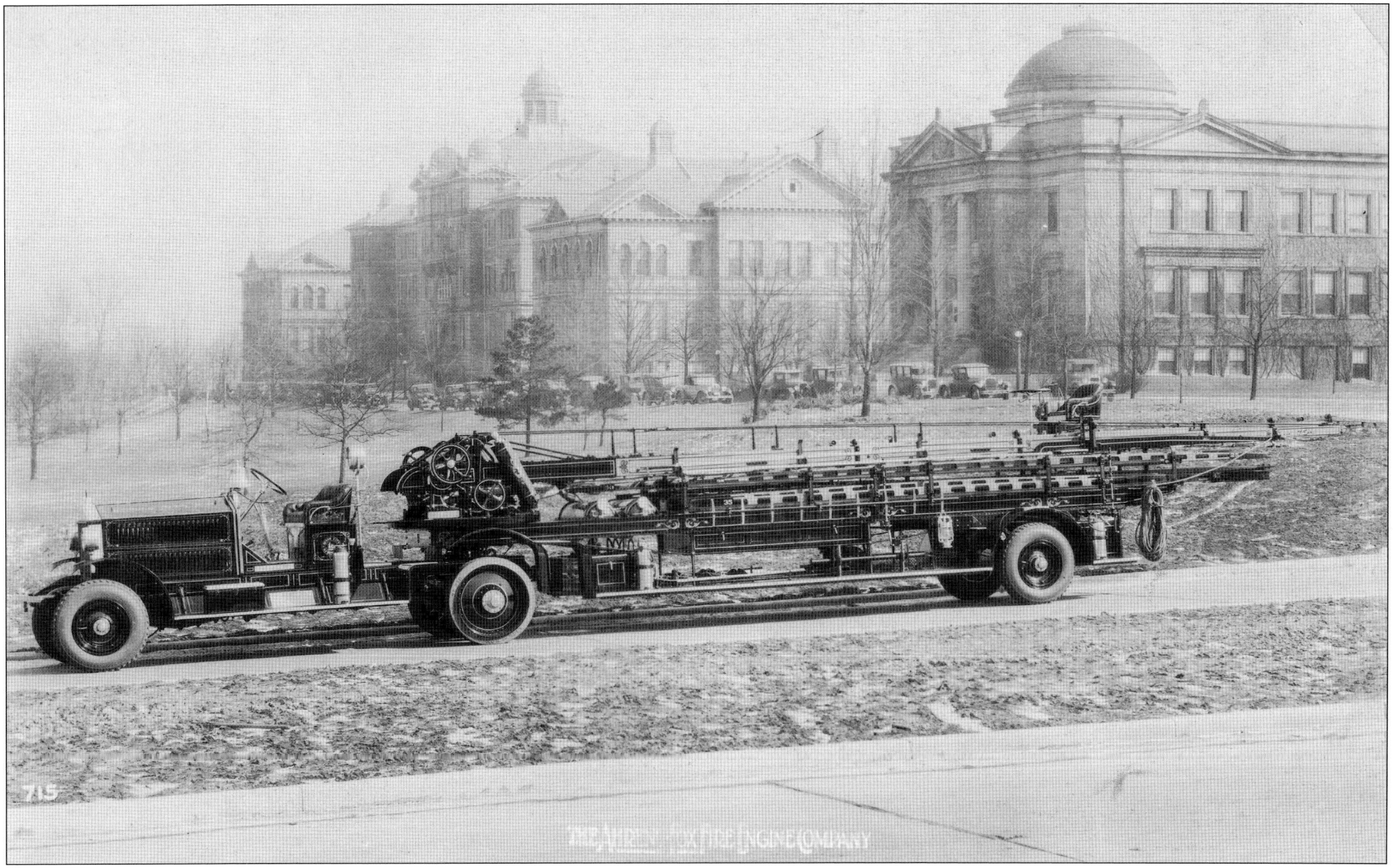

1926 Tractor-Trailer Aerial—A unique feature of Ahrens-Fox's tractor-drawn aerial truck was its pneumatic ladder hoist. The two-section 75- or 85-foot wood aerial ladder was raised and lowered by compressed air rather than by the coil springs favored by Ahrens-Fox's principal competitors—American-LaFrance and Seagrave—on their aerial trucks. Invented by Chief Edward F. Dahill of the New Bedford, Massachusetts, Fire Department, the *Dahill Air Hoist* was used on most of the more than 70 aerial trucks Ahrens-Fox built between 1923 and 1940. Ocean City, New Jersey, received this 75-foot *Model 75-6-1,* #2015, in 1926. All Ahrens-Fox aerials were powered by six-cylinder engines. Note the disc wheels, pneumatic tires front and rear, and the load-bearing hard rubber tires on the rear axle of the tractor. *MLC*

1926 Model PS4 Piston Pumper #738—Passaic, New Jersey, boasted a large fleet of Ahrens-Fox piston pumpers—most of which were expertly photographed in their prime by a young Dick Adelman who grew up in the area. Dick photographed Passaic Engine 2's 1,300-gpm PS4 on the street at a fire in the early 1950s. Note the shop-installed windshield, the odd warning light in front of the fuel tank, and the curved rear handrail. *Richard M. Adelman*

1926 Model NS3—In full turnout gear and ready to roll, members of the Colorado Springs (Colorado) Fire Department pose with Engine No. 1's spanking new Ahrens-Fox *Model NS3* piston pumper #1662 in this photo taken on May 6, 1926. Note the steel disc wheels. Among the most popular of all Ahrens-Fox piston pumpers, the 1,000-gpm *NS* was powered by a 110 horsepower six-cylinder engine with 4½" bore and 6" stroke. *MLC*

1926 Model JS-44 Service Ladder Truck—Sanford, Maine, was the original purchaser of this 1926 *Model JS-44* Double-Bank City Service Ladder Truck which, following its retirement, was on display at the New England Fire & History Museum in Brewster, Massachusetts, for many years. Lettered *Fernwood* on its gabled hood, Reg. #1258 resides today in the new fire apparatus museum in Circleville, New York. Of special interest on this nicely preserved old ladder truck is the rare four-lens *Whiting* warning light behind the driver's seat. *WMC*

1926 Model MX14 Combination Hose Wagon—Some Ahrens-Foxes had extended, multiple fire service careers. This 1926 *Model MX14* hose and turret car was originally purchased by Morristown, New Jersey. Reg. #1309 was later sold to Fort Lee, New Jersey, where it soldiered on for many more years. Note the fixed turret pipe in the hose bed, the retrofitted windshield and the flashing red light on the radiator. The bell has been relocated from the dash to a new location above the front bumper. *MLC*

New York Special –*"Engine 65 of the New York Fire Department is the world's most powerful fire engine,"* Ahrens-Fox boasted in its advertising following the delivery of this special high pressure pumping engine to the FDNY in 1927. The 1,000 gallon-per-minute *Model HP2, #774,* was assigned to busy Engine Company 65 in midtown Manhattan. New York's "*Super Fire Engine*" was designed to deliver 1,000 gallons of water per minute at a pressure of 160 pounds, supplying two hose streams to the top of the Woolworth Building nearly 800 feet above street level. Ten years earlier, another FDNY Ahrens-Fox piston pumper, an MK2, also pumped through a six-inch standpipe to the top of the Woolworth Tower—at the time America's tallest skyscraper. Engine 65's famous rig was powered by a 140 horsepower six-cylinder engine coupled to a Twin Triple six piston pump. *MLC*

Owner's Pride—Among old fire engine buffs, a *"muster"* is a gathering of antique and vintage fire engines—sort of like a classic or antique car show but with a lot more red paint! For many years one of the largest annual antique fire apparatus musters in the U.S. was hosted by SPAAMFAA's Great Lakes Chapter at world-famous Greenfield Village in Dearborn, Michigan. Here John Callaghan of Grand Blanc, Michigan, pilots his beautifully restored 1927 Ahrens-Fox MS4, #1732, ex-Kansas City, Missouri, through the village at a Greenfield Muster in the mid-1970s. *Henry Ford Museum*

Kansas City Classic—Kansas City, Missouri, purchased a large fleet of Ahrens-Fox fire apparatus in 1927. The order for no fewer than 21 pieces of apparatus included ten MS4 1,000 gpm piston pumpers; six NS14 combination hose and chemical cars, four 75-foot tractor-drawn aerials and an NS44 city service hook and ladder truck. Many of these rigs were retired and sold off in the late 1960s. In 1970, collector Andy Henderson of Weedsport, New York, bought one of the aerials (#2026) which he and two sons drove directly from Kansas City to Weedsport just in time for the 1970 SPAAMFAA muster in Syracuse. The transcontinental trek took four days, and more than 540 gallons of gasoline. *JFS*

One Man's Generosity—In one of the most inspiring acts of civic-mindedness we've ever encountered, a wealthy industrialist in 1927 purchased *five* brand-new Ahrens-Foxes for the Consolidated Fire Company in his home town of Bristol, Pennsylvania. They included, from left to right, with a 1928 Buick Roadster Chief's Car, a 600-gpm *Model UN6G* rotary gear Quadruple Combination, a 750-gpm *Model MS4* piston pumper, a *C-14* Combination Hose and Chemical Car, an *NS2* 1,000-gpm piston pumper, and a second *C-14* Hose and Chemical Combination with turret nozzle. *MLC*

1928 Combination Hose & Booster Car—Ahrens-Fox phased in some subtle design changes during 1928. Flat-top fenders were replaced with new stamped steel "crowned" fenders. Thin-profile *Ryanlites* replaced the Vesta drum headlights used earlier and all models got a new double-bar front bumper. These features are all visible on this 1928 *Model GN-10-4 Combination Hose and Booster Car #5010* for Somerville, Massachusetts, with disc wheels and solid-rubber tires. *MLC*

Matched Pair Of `28s—Pleasingly posed on the ramp in front of the firehouse in Green Bay, Wisconsin, are a matched pair of gable-hood Ahrens-Foxes. Delivered in 1928, they include a *Model NS44* Double-Bank City Service Ladder Truck (#3330) with chemical equipment and fixed turret nozzle, and *NS4* piston pumper #3337. Both sport *Ryanlite* headlights and the new crowned fenders and double-bar front bumper. Note the preconnected 2½" hard suction hoses on the pumper. *JFS/MLC*

1928 Model NS3 Piston Pumper—Cleveland, Ohio, fire buff Paul Nelson took this fine profile shot of a classic Ahrens-Fox piston pumper late in its fire service career. Lynn, Massachusetts' 1928 *Model NS3, #3324* with 1,000 gallon-per-minute four-piston pump sports the extra-thin *Ryanlite* headlights found on Foxes in the late 1920s. The 998 cubic inch, 110-horsepower six-cylinder Ahrens-Fox motor with its individual cylinder jugs is visible through the open engine compartment. Note the chemical tank and tall, old-style turret pipe atop the hose bed. *JFS*

Ahrens-Fox *Rotarystyle* Pumper—Ahrens-Fox was justly famed for its distinctive, highly dependable front-mount piston pumpers. But the rotary gear and centrifugal pumpers offered by competitors were gaining in favor with the fire service. To meet this growing competition, Ahrens-Fox in 1925 introduced a conventionally styled pumper with a midship-mounted rotary gear pump. The six-cylinder Ahrens-Fox *Rotarystyle* was available in 500-, 600- and 750-gpm sizes. Beverly Hills, California, purchased a pair of 750-gpm *Model GN-80-4* rotary gear pumpers in 1928—one of which got a frame-up restoration in the 1980s, as pictured here. *Beverly Hills Fire Department*

1928 Ahrens-Fox *Satellite* Pumper—Charles H. Fox was granted a patent for his company's improved *Double-Seal* rotary gear pump. Expanded to a full line of pumpers and quads by 1928, Ahrens-Fox's rotary-gear models got their own distinctive styling with a new softly contoured radiator, hood and cowl. The Toledo suburb of Maumee, Ohio, still owns this 1928 *Model GC-6-4 Satellite 600-gpm pumper, #5006,* which was photographed by John F. Sytsma on a slushy winter day. *JFS/MLC*

Ahrens-Fox *Skirmisher* Quad—In 1928, Ahrens-Fox expanded its rotary-gear product line with the addition of a quadruple combination called the *Skirmisher.* A combination pumper/ladder truck, the "quad" combined four firefighting functions on one chassis: pump, booster system, hose bed and full complement of ground ladders. Designed for towns and small cities, the *Skirmisher* was built in 600- and 750-gpm versions powered by a Waukesha six-cylinder engine. Glendale, Ohio, still owns this 600-gpm 1929 *Model UW-60-4 Skirmisher. JFS*

Below: 1928 *Model NS-444 Service Ladder Truck, #5017,* Pawcatuck, Connecticut. *MLC*

Above—1928 Ahrens-Fox *Skirmisher* Rotary Gear Quadruple Combination. *MLC*

Below—1928 Ahrens-Fox *Satellite* Triple Combination Pumper with rotary gear pump and booster equipment. *MLC*

Cannonball Row—New Orleans upgraded its fire department with the purchase of ten Ahrens-Foxes in 1927 and 1928. Here are all ten of them lined up on a downtown street in June of 1928. This impressive firefighting fleet includes eight *Model "M"* 750-gpm piston pumpers, one *Model HP* piston pumper and a city service hook and ladder truck, which is barely visible next to the fifth pumper from the left. *MLC*

1929 *Standard Line* Quadruple Combinations—As the 1920s drew to a close, Ahrens-Fox was building two distinct types of quads. **Above:** 1929 *Model UM4, #1764,* with front-mount piston pump for Cheviot, Ohio. **Below:** 1929 *Standard Line* quad with midship type rotary gear pump in service in Wilmington, Delaware. Note the fire company number painted on the hard suction hose. *MLC*

Canadian Ahrens-Fox—In the mid-1920s, Ahrens-Fox signed an agreement with Bickle Fire Engines Ltd. of Woodstock, Ontario, under which Bickle would build Ahrens-Fox fire apparatus for Canadian fire departments. Four *Bickle-Ahrens-Fox* piston pumpers were built under this arrangement—an NS2 for Kingston, Ontario, in 1924 and three NS4's for Hamilton, Ontario, in 1926, 1928 and 1930. In the mid-1950s two of Hamilton's 800 Imperial gpm (1,000 US) pumpers were converted into *Deluge Pumpers* by the H.F.D. shops. Deluge Pumper No. 16, shown here, eventually wound up in Spokane, Washington. *Ken Buchanan*

A Face Only A Fire Fighter (Or Fire Buff) Could Love—Crowned front fenders and gable radiator indicate this is a 1929-1931 piston pumper. The short length of 2½" hose connected to the hydrant leads into a reducer in the *single* six-inch front suction intake. The model and location are unknown. We have the urge to just reach *into* this photo and spin that hydrant wrench! *MLC*

1929 Ahrens-Fox 85-6-1 Aerial—Ahrens-Fox's pleasing new styling extended to the company's big tractor-trailer aerial ladder trucks. The handsome new front-end ensemble included a new softly-rounded hood, radiator and cowl. New Orleans, Louisiana, purchased two of these 85-foot aerials, Reg. #2044 and #2055, in 1929. This is one of them late in its fire service career. Note the hard rubber tires on the rear axle of the tractor, replacement sealed-beam headlights, metal ground ladders and the two compressed air tanks for the Dahill air hoist. *NOFD/MLC*

Ahrens-Fox Tower Aerial—Ahrens-Fox extensively redesigned its tractor-drawn aerial trucks in the early 1930s, strengthening the upper section of the ladder to permit it to be used as a water tower. **Above:** 1930 *Model 85-6-1* 85-foot Tower Aerial, #2054, for Dayton, Ohio. Note the whitewall tires! **Below:** Unidentified city service ladder truck showing handsome late 1920s/ early 1930s front-end ensemble. *MLC*

1930 Model PS-14 Combination Hose Car—The only Ahrens-Fox fire engines ever purchased by the Los Angeles City Fire Department were five of these big Combination Hose and Chemical Cars delivered in 1930, Reg. #2306–2310. Features of note include the oversized hose body, single 60-gallon soda and acid chemical tank, side-mounted spare tire and top-mounted chemical hose reel. Note also the small carbon tetrachloride fire extinguisher mounted below the seat riser. These jobs also sported non-standard headlights. *MLC*

1931 Quadruple Combination—One of the most spectacular Ahrens-Foxes still on the road today has to be Tarentum, Pennsylvania's beautifully restored 1931 *Model UNT-4 Quad,* #3404. This stunning combination pumper and city service ladder truck has a 1,000-gpm piston pump. This would be the final year for the gable-style hood first seen in 1919. Tarentum's pride and joy was photographed by the late Roger O. Jackson in 1975. *WMC*

1932 Schacht Model "V" 500 GPM Pumper—In 1930, as the Depression deepened, Ahrens-Fox introduced a new line of economical small-town fire apparatus built on light- duty commercial truck chassis purchased from the LeBlond-Schacht Truck Company of Cincinnati, Ohio. Produced through 1935, the Model "V" ultimately sank Ahrens-Fox. Deeply in debt for these chassis, Ahrens-Fox was forced to merge with LeBlond-Schacht in 1936, at which time the company moved from its original home on Colerain Avenue into the LeBlond-Schacht plant. Things would never be the same. Originally delivered to Amherst, Ohio, this 1932 Model "V," #7031, was later acquired by North End, New York. *JFS*

1933 Model AHP Piston Pumper—The most powerful pumpers Ahrens-Fox ever built were a pair of truly awesome *Model AHP* twin-triple piston pumpers delivered to New York City (FDNY) in 1933. Reg. #4003 and 4004 could deliver a brick-crushing 1,500 gallons of water per minute. Note the cast-steel wheels shod with solid rubber tires, fixed turret nozzle, and overhead-mounted pompier ladders. **Below:** In 1934 the FDNY received two very similar-looking 1,000-gpm *Model NT2s* with pneumatic tires. *MLC*

1933 Model CT4 Piston Pumper—Before the legendary *Model HT* came the *CT, BT* and *IT.* The 1933–34 *Model CT* had a six-piston 900-gpm pump, the later *BT* a 1,250-gpm Twin-Triple pump and the *IT* a 750-gpm six-piston pump. York, Pennsylvania, received this imposing *Model CT4* (#4006) in 1933. Note the chrome-plated plank-style front bumper, Budd steel wheels, factory-installed windshield with dual electric wipers, and the chrome-plated ventilation doors on the hood. *MLC*

1934 Model VLB Hose Car—At the depths of the Depression, Hoboken, New Jersey, purchased a substantial fleet of Ahrens-Fox fire apparatus. This big order—which undoubtedly helped Ahrens-Fox weather the storm—included five *Model CT4* piston pumpers and three of these Schacht-chassis *Model VLB Combination Hose and Booster Cars*. Ahrens-Fox historian John F. Sytsma photographed this one at an antique fire apparatus muster in 1974. *JFS*

1935 Model VC Centrifugal Pumper—In 1935 Ahrens-Fox replaced the Schacht-chassis Model "V" with an entirely new series of mid-sized fire engines aimed at smaller fire departments. Produced through 1938, this pleasingly styled series was available with either a Hale centrifugal pump or Northern rotary gear pump. The front end featured a distinctive shield-shaped chrome-plated grille. This is a rotary-gear *Model ER #8011* delivered to Mechanicsburg, Pennsylvania in 1935. *MLC*

1936 Model BT 1,250 GPM Piston Pumper—Predecessor to the classic *HT,* the big-bore 1,250 gallon-per-minute *Model BT* was produced from 1935 to 1939. Just nine were built. The only Ahrens-Fox piston pumper ever purchased by the Boston Fire Department was this 1936 Model BT #4014. **Below:** Arguably one of the most beautiful fire engines ever made in America was Lewiston, Pennsylvania's stunning cream-over-dark-blue 1938 *Model BT* #4022. *MLC*

1936 Model CR Centrifugal Pumper—Justly famous for its front-mounted piston pumps, Ahrens-Fox also began offering pumpers equipped with rotary gear and centrifugal pumps in the mid-1930s, while continuing production of its famed piston types. The Liberty Hose Company of Martins Ferry, Ohio, ordered this one-of-a-kind 1,250 gallon-per-minute rotary gear pumper, #4012, in 1936. Note the massive grille and bumper and knobby all-weather tires. *MLC*

1937 Model XCT Sedan-Bodied Piston Pumper—Pawtucket, Rhode Island, ordered the only Ahrens-Fox piston pumper ever built with a fully-enclosed van-type body. The one-off limousine style body was subcontracted to a local truck body builder, the *Kelly Body Co.* of Cincinnati. Bearing Serial #4021, this unique 750 gallon-per-minute pumper now resides in the antique fire apparatus museum in Circleville, New York. *MLC*

1937 Model VC Sedan-Bodied Pumper—The mid-sized *Model VC/VR* line was offered in a full range of body styles, from wide-open doorless cabs to fully enclosed van-type bodies. Painted black with red wheels, this 1937 Ahrens-Fox *Model VC* 500-gpm pumper with fully enclosed sedan style body was built for the West End Fire Company No. 7 of Pottsville, Pennsylvania. Ahrens-Fox subcontracted its fully enclosed bodies to the Kelly Body Co. of Cincinnati. *MLC*

1937 Model CTU Quadruple Combination, #4019—Ahrens-Fox built only *eight* Quadruple Combinations, or "quads" with front-mounted piston pumps in the decade between 1927 and 1937. This was the last. Built for the Borough of Ho-Ho-Kus, New Jersey, the combination pumper and ladder truck was equipped with a 750 gallon-per-minute four-piston pump. Note the new style "plank" bumper. This beauty got away from Ho-Ho-Kus, but in one of those happy-ending stories was later bought back and restored to its original appearance. *Henry Schmults Jr.*

1937 Model CTU Quadruple Combination—Here's what Ho-Ho-Kus, New Jersey's 1937 Ahrens-Fox Quad looked like the day it was delivered more than 70 years ago. That's Fire Chief Dave Miller behind the wheel. Note the open equipment compartment behind the front seat and the angled hose playout chutes at the end of the divided hose bed, which kept the 2½" hose from becoming entangled in the ladders when laying lines. *Ho-Ho-Kus Fire Department*

The First Model HT—Introduced in 1937, the 1,000 gallon-per-minute *Model HT* marked the pinnacle of Ahrens-Fox piston pumper design. The new HT, powered by the six-cylinder Hercules HXE engine, had semi-streamlined styling with skirted front and rear fenders, curved body sheet metal behind the seat and a one-piece, fold-down windshield. A total of 67 HTs were built before the last one was delivered 15 years later. Collector John Gambs of Lafayette, Indiana, now owns the first HT delivered, #3420, which went to South St. Paul, Minnesota, in late 1937. *John R. Gambs*

FDNY Closed-Cab HT—A single customer—New York City—purchased nearly *half* the Model HTs Ahrens-Fox produced between 1937 and 1952. Deck turret cranked skyward, this is one of the twenty 1,000-gpm pumpers with three-man closed cabs delivered in 1938. Two more `38s followed for the 1939-1940 New York World's Fair and another six were delivered in 1939. Note the siren mounted atop the vacuum chamber in front of the trademark spherical air dome, the rear-mounted spare tire and the windshield for the men riding the rear step. These distinctive fire engines served the FDNY into the 1960s. Just one survives, in the New York State Museum in Albany. *MLC*

1938 Model HT Piston Pumper Order—The New York City Fire Department (FDNY) placed one of the largest single orders ever received by Ahrens-Fox. Dated January 7, 1938, the order was for 20 special *Model HT* piston pumpers with three-man closed cabs. Amazingly, this big order was completed in just over five months with the last one delivered on May 21. The FDNY ordered six more HTs in 1939, plus two for the New York World's Fair, for a grand total of 28. *WMC*

New York World's Fair—New York City ordered a special fleet of five Ahrens-Foxes to protect the 1939-1940 New York World's Fair. Delivered in 1938, they included two 1000 gpm *Model HT* piston pumpers (bottom left) and three 500 gpm Model SC-500 pumpers (above), also with closed cabs on Schacht commercial truck chassis. After the fair closed all five went to regular FDNY fire companies. Note the Stature of Liberty emblem on the door of the HT. *MLC*

Closed Cab Styles—Closed cabs swept the fire apparatus industry in the latter half of the 1930s, and Ahrens-Fox was in the vanguard of this safety design trend. **Above:** Note the rakish coupe-style cab on this 1938 *Model HC* centrifugal pumper #9051 for Stamford, Connecticut. **Below:** Similarly styled 1938 *Model EC* #9058 for Larchmont, New York.

1938 Model HC Centrifugal Pumper—Ahrens-Fox, it seemed, would build *anything* for a loyal customer like its hometown Cincinnati Fire Department. In 1938 the company built three of these special *Model HC* pumpers (#9055, 9056 and 9057) for the C.F.D. What made them truly unique was that they were equipped with 1,000 gallon-per-minute centrifugal pumps purchased from rival Seagrave of Columbus, Ohio. Other features of note include the painted radiator shell, the Buckeye *Roto-Ray* warning light behind the coupe style cab and the rear step windshield complete with New York-style "subway-strap" hangers. *MLC*

1938 Model VR Rotary Gear Pumper—This classic "ramp shot" taken by John F. Sytsma in 1972 captures nicely the pleasing front-end styling of the mid-sized Ahrens-Fox VC/VR series produced between 1935 and 1938. Note the dipped front bumper, tall tapered grille and cowl-contoured one-piece windshield. Dover Township, Ohio, where this 500-gpm rotary gear pumper was originally delivered, was later renamed Westlake. Note the side-mounted bell. The post-mounted beacon is a retrofit. *MLC*

Model EC Triplets—Des Moines, Iowa, received this trio of Ahrens-Fox Model EC 750 gpm centrifugal pumpers the day before Christmas of 1938. Purchased at a cost of $8,165 each, they bore serial numbers 9065, 9066 and 9067. When delivered, they were lettered for Engines 1, 2 and 3—but were actually assigned to Engines 3, 4 and 5. *Bob Stinson Collection*

Streamlined Styling— Some Ahrens-Fox piston pumpers built in the late 1930s left the Cincinnati plant with some fetching custom coachwork. Here are two fine examples. **Above:** 1939 *Model BT* 1,250-gpm pumper #4024 with closed coupe-style cab for Shelton, Connecticut. **Below:** Ultra-streamlined, fully enclosed 1939 *HT* 1,000-gpm pumper #3447 for Ocean Grove, New Jersey. *MLC*

1939 Model BJQ-70 Quintuple Combination—This humungous piece of fire apparatus was the largest—and most expensive—ever made by Ahrens-Fox. Designed and built for the Fairview Fire Department of Greenburgh, New York, the special *Model BJQ* (#2068) was equipped with a 1,000-gpm piston pump and 70-foot Ahrens-Fox *Bascule* aerial ladder. Price was a staggering (for the time) $20,000. This firefighting juggernaut was retired and sold to a private collector in the mid-1970s. *MLC*

1939 Schacht Chassis—After a four-year hiatus, Ahrens-Fox again began building on Schacht chassis. For 1939, Schacht introduced this rather bland design which Ahrens-Fox used in 1939-40. **Above:** 1939 *Model EC* with coupe-style cab (#9076) for Morgan City, Louisiana. **Below:** 1939 *Model VC* (#9079) built for Elfinwild-Glenshaw Pennsylvania. *MLC/WMC*

Long-Wheelbase Models— The Ahrens-Fox model lineup continued to include long-wheelbase city service ladder trucks and quadruple combinations—although these were now few and far between. **Above:** *Model 4VD* city service ladder truck, #8039 built for Salem, New Jersey, in 1940. **Below:** 1939 *Model VRU Quad #8034* for Pascoag, Rhode Island. *MLC*

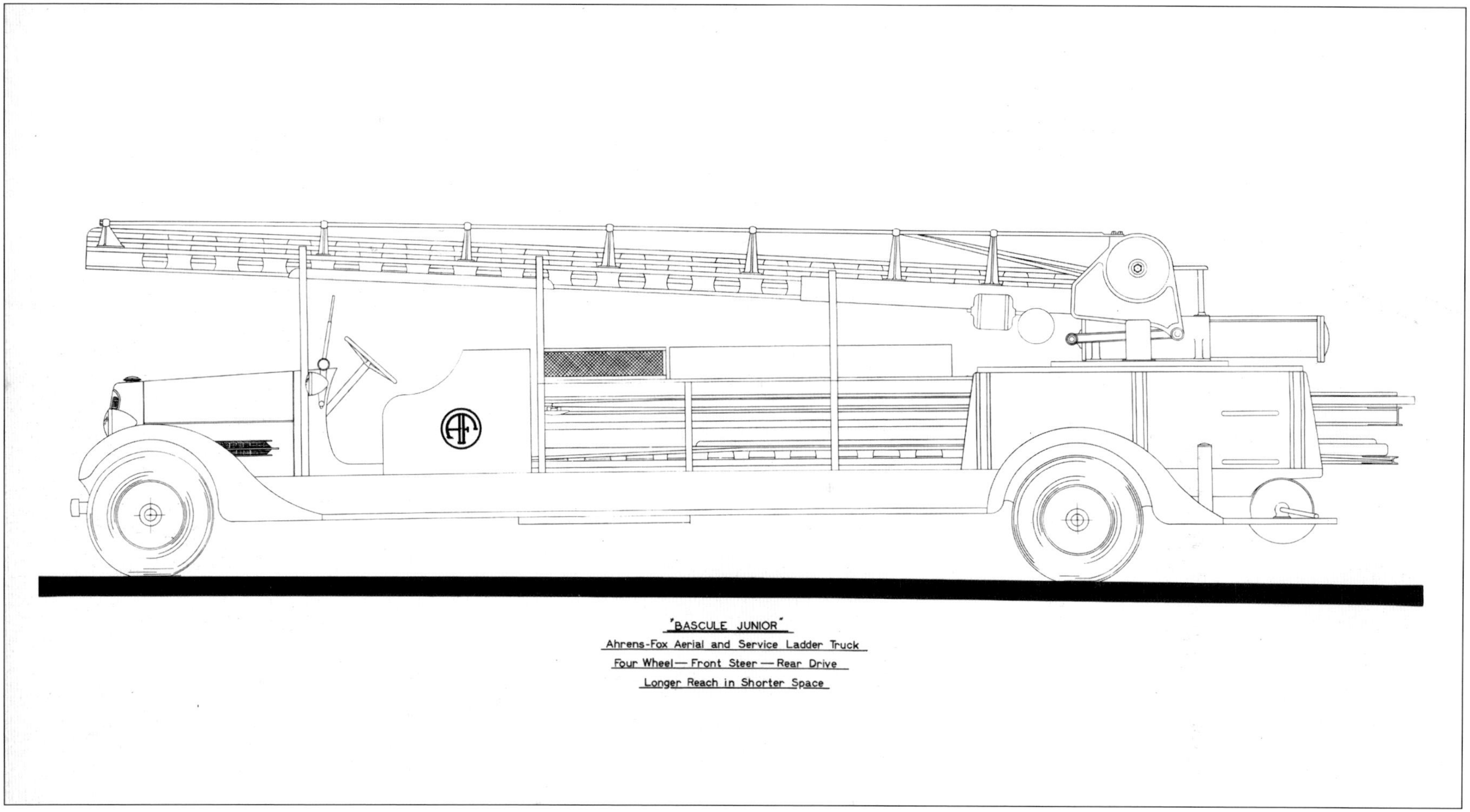

Ahrens-Fox "Bascule Junior" Rearmount Aerial—This rearmount aerial truck design was *two decades* ahead of its time. The circa 1938 design proposal shows a three-section metal aerial ladder mounted at the rear of a conventional Ahrens-Fox V-Series city service ladder truck chassis. Maxim introduced the U.S. fire apparatus industry's first custom-chassis rearmount aerial in 1954, but the rear-mount concept did not really catch on in the U.S. until the late 1960s. *MLC*

1940 Model EC Pumpers—LeBlond-Schacht's *Ahrens-Fox Division* utilized the new style Schacht commercial truck chassis in 1939 and 1940. Although Schacht ceased truck production in late 1939, at least one more Ahrens-Fox of this style was assembled using leftover sheet metal stock in 1946. The new Schacht featured a tall, narrow screen grille. Lima, Ohio, received three *Model EC* pumpers of this style in 1940. Bearing factory serial numbers 9084, 9085 and 9086, they had 750 gallon-per-minute centrifugal pumps and coupe-style cabs. Note the roof-mounted sirens. *MLC*

1940 Model W-85 Tractor-Trailer Aerial—Ahrens-Fox delivered its last aerial trucks in 1940 concluding a 17-year production run that began in 1923. The New York City Fire Department (FDNY) received four of these big 85-foot tractor-trailer aerials with two-section, spring-raised wood aerial ladders. Note the wig-wag warning light atop the massive cast-aluminum radiator shell and the two pompier (scaling) ladders carried on the trailer frame. This one was photographed in the plant yard prior to delivery to Gotham. San Francisco purchased two Ahrens-Fox 85' aerials of this style in 1939. *MLC*

***War Baby*—1942 Model HT Piston Pumper—**This is the only fire engine Ahrens-Fox built during the Second World War. With parent company machine tool builder LeBlond-Schacht heavily committed to war production, the company's Ahrens-Fox Division ceased motor fire apparatus production in 1940. This 1,000-gpm *Model HT* piston pumper (#3457) was assembled out of existing parts stocks, with hard-to-obtain Defense Department permission, for Harrison, New Jersey—home to some vital defense industries. The Harrison Fire Department received an identical HT when Ahrens-Fox resumed peacetime fire apparatus production in early 1946. *MLC*

The Classic Model HT—Although the centrifugal pump now dominated the industry, Ahrens-Fox continued to offer its venerable front-mounted piston pump, but there were fewer and fewer takers. **Above:** A twin to the one delivered in 1942, Harrison, New Jersey, got HT #3458 (with wartime painted siren) in early 1946. **Below:** 1947 HT #3466 for Chester, Pennsylvania. Note the chrome-plated rooster mascot atop the air chamber. *MLC*

1947 Model HT #3462—New Bedford, Massachusetts, ordered two Ahrens-Fox *HT* piston pumpers after the war. This is #3462 which was delivered in early 1947. A twin, #3475, was delivered the following year. The 1,000-gpm HT was powered by the 210 horsepower, 935 cubic inch *Hercules HXE* six-cylinder engine of 5¾" bore and 6" stroke. Wheelbase was 202 inches. With the driver sitting virtually amidships, the HT was a handful to drive. *MLC*

Postwar Styling—In 1947, Ahrens-Fox introduced an all-new conventional custom fire apparatus design styled by Richard C. (Curt) Nepper who had joined the company in the late 1920s. Hallmarks of this pleasing design, which remained in production through 1955, included a softy-contoured front-end with two-section grille and big squared-off front fenders. This *Model VC* pumper, #9107, was delivered to the New Market Fire Company in Piscataway, New Jersey, in 1948. *MLC*

1948 Model VC Special—Ahrens-Fox's new long-hood conventional was available in a wide range of body styles—from the standard open-cab type with or without doors on up to this long-wheelbase version with four-door sedan type cab. This large pumper, a *Model VC Special, #9103,* was built for Hollsopple, Pennsylvania. Note the dual air horns on the hood ahead of the windshield and the siren mounted on the cab roof. For such a big rig, all-around visibility from the cab was excellent. *MLC*

1948 Closed-Cab Model HT—New Orleans received four Ahrens-Fox *Model HT* piston pumpers with three-man closed cabs in 1948. Fresh off the assembly line, this one was photographed outside the Cleveland Automatic Machine Co. plant on Beech Street in the Cincinnati suburb of Norwood prior to shipment to the Crescent City. Note the *Buckeye Roto-Ray* warning light atop the stubby vacuum tank and the extra length of hard suction hose. These closed-cab beauties remained in front-line service into the 1970s. *MLC*

Ready to Ship—This photograph was taken in the Cleveland Automatic Machine Co. plant yard in Norwood in the winter of 1947-48. In the foreground is the first of the four closed-cab *Model HT* piston pumpers for New Orleans, Louisiana. Visible in the background is another unidentified HT—this one with an open cab. Note the different style front bumpers. Ready for its delivery trip, the HT in the foreground sports a 1947 Ohio dealer license plate. *MLC*

1949 Model FC Special—From the time they were introduced in 1947, Ahrens-Fox's new postwar custom conventionals far outsold the company's famed piston pumpers. West Mifflin, Pennsylvania, received two of these *Model FC Special* centrifugal pumpers with open semi-cabs—#9116 in 1948 and a twin, #9150, delivered the following year. Note the compartmented rear fenders, the *Ahrens-Fox* nameplate below the cab door, the Mars warning light above the windshield and the two demountable floodlights at the front of the hose bed. *MLC*

1949 Model HT With Cab Doors—Eclipsed by the far less complex centrifugal, the piston pump was by now an anachronism, a hold-over from the past. By the late 1940s sales of Ahrens-Fox piston pumpers had plummeted to only a few a year. Nonetheless, in 1948 Turtle Creek, Pennsylvania, ordered this *Model HT* with open semi-cab cab, which was delivered in 1949. Painted a striking *Turtle Green*, the late Roger O. Jackson photographed #3482 in the early 1970s. The Beacon-Ray light above the wide two-piece windshield and West Coast mirrors were retrofits. *WMC/Roger O. Jackson*

1950 Model ICU Quad—The small Ohio city of Mount Healthy had the distinction of running *two* Ahrens-Foxes—this impressive long-wheelbase quad and a 1938 *Model VR.* Equipped with a 750-gpm centrifugal pump, #9153 has an open semi-cab and enclosed ladder bed. Note the dual booster hose reels behind the front seat and the post-mounted Federal Beacon Ray warning light. *JFS/WMC*

Last Piston Pumper—Forty years after the first double-dome *Model "A" Continental* was delivered to Rockford, Illinois, in early 1912, Ahrens-Fox delivered its last piston pumper to Tarrytown, New York, in the spring of 1952. The 1,000-gpm Model HT, #3488, is still owned by the Hope Hose Company of Tarrytown. This milestone piece of motor fire apparatus was photographed at one of the big antique fire apparatus musters held at the Kenisco Dam Plaza in Valhalla, New York, in the mid-1970s. *MLC*

1954 FC Special—In 1953, owner Walter (Duke) Walkenhorst subcontracted Ahrens-Fox fire apparatus production to the *C. D. Beck Company*, a builder of transit buses located in Sidney, Ohio. For the next three years Ahrens-Fox fire engines were built alongside Beck's big intercity buses. A longtime Ahrens-Fox customer, Salt Lake City, Utah, ordered this 1,000-gpm pumper with rear entrance canopy cab (#9163) in 1954. The Salt Lake Fire Department received a similar-looking Ahrens-Fox (#9096) in 1947. *MLC*

1956 Cab-Forward Pumper—In early 1956 Ahrens-Fox surprised the rest of the industry with its introduction of a state-of-the-art *cab forward* pumper. Utilizing the same front-end sheet metal used on the company's buses, the Beck-built *Model FCB* #2001 was powered by a Waukesha six-cylinder engine and equipped with a Hale 1,000-gpm centrifugal pump. Completed in late 1955, the prototype pumper—a factory demonstrator with open semi-cab—was later sold to East Rutherford, New Jersey. Note the fifties-style "V" emblem on the smooth cab face. *MLC*

Last Of The Line—The last fire engine to bear the legendary Ahrens-Fox name was this Beck-built *Model ECB* 750-gpm pumper, #2006, ordered by Dumont, New Jersey, in 1957. But when Dumont learned that Ahrens-Fox was going out of business they promptly cancelled their order. Northern Hills, Ohio, eventually purchased the "orphan" pumper. The excellent Ahrens-Fox cab-forward design was quickly appropriated by Mack—which had acquired C.D. Beck—and lived on for another 12 years as Mack's highly successful C-Series fire apparatus. *JFS*

Color Gallery

Only known color photo of a double-dome Ahrens-Fox: 1915 *Model AC Continental* 750-gpm Piston Pumper #557, Danvers, Massachusetts. *Courtesy of Leonard W. Williams*

1917 Ahrens-Fox *Model M-K-2* Six-Cylinder 800-gpm Piston Pumper, #828, ex-Concord, Massachusetts, restored by Nelson Barber. *Dan Jasina Photo*

1917 Ahrens-Fox *M-17 Tractor* #842 attached to 1896 Hale 55' Water Tower, restored and owned by Kenneth W. Menke. *Ken Burrows Photo*

1920 Ahrens-Fox *Model I-K-4* Four-Cylinder 750-gpm Piston Pumper #772, ex-Danville, Kentucky. Owned by Jerry Voisinet. *Dan Jasina Photo*

1921 Ahrens-Fox *Model T-3* Piston Pumper, #1002 with five-piston 500-gpm pump, Batesville, Indiana. *James Tharp Photo*

1922 Ahrens-Fox 750-gpm *Model I-M-2* Piston Pumper, #1061, ex-Chicago Fire Department, owned by Keith Franz. *Walt McCall*

1922 Ahrens-Fox *Model K-J-S-4* Four-Cylinder 750-gpm Piston Pumper, #1114, Ex-Baltimore, Maryland, Maryland Fire Museum, Baltimore. *Chuck Madderom Photo*

1922 Ahrens-Fox *Model J-13* High Pressure Hose Car #1204, one of two built for Detroit Fire Department, owned by Mike Adams, Burt, Michigan. *Walt McCall Photo*

1923 Ahrens-Fox *Model K-17* Tractor-Drawn City Service Hook & Ladder Truck, #1110. Ex-Cincinnati, Ohio, and Batesville, Indiana, now owned by Jimmie Dobson, Bay City, Michigan. *Walt McCall Photo*

1924 Ahrens-Fox *Model 75-6-1* Tractor-Drawn 75' Aerial, #2004, ex-Kokomo, Indiana. Owned by Mike Adams. *Walt McCall Photo*

1926 Canadian-built *Bickle-Ahrens-Fox N-S-4* #1671, 840 imperial gpm, ex-Hamilton, Ontario, now maintained by Box 43 Association of Hamilton. *Ken Buchanan Photo*

1927 Ahrens-Fox 750-gpm *Model M-X-4* Piston Pumper #1314 ex-Bristol, Connecticut, owned by Nelson Barber. *Walt McCall Photo*

1928 Ahrens-Fox *N-S-4* 1,000-gpm Piston Pumper, #3326, ex-Newport, Kentucky, owned by Roger Engvall. *Walt McCall Photo*

1928 Ahrens-Fox *Model GC-60-4 Rotarystyle* Rotary Gear Pumper, #5006, still owned by Maumee Fire Department, Maumee, Ohio. *Walt McCall Photo*

1931 Ahrens-Fox 750-gpm *Model M-X-2* Piston Pumper, #1318, ex-Detroit Fire Department, owned by Leonard W. Williams. *Walt McCall Photo*

1931 Ahrens-Fox *Model U-N-T-4* Quadruple Combination, #3404, 750-gpm piston pump, owned by Tarentum, Pennsylvania. *Shaun P. Ryan Photo*

1931 Ahrens-Fox *Model N-T-4* Piston Pumper, #3403, ex-North Attleborough, Massachusetts, owned by Bob Cauger. *Dan Jasina Photo*

1931 Ahrens-Fox *Model 85-6-1* Tractor-Drawn 85' Aerial Ladder Truck, #2056. Ex-Harrison and Wayne, New Jersey. Owned by Andy Henderson. *Walt McCall Photo*

1931 Ahrens-Fox *Model N-S-444* City Service Ladder Truck, #3402, ex-Worcester, Massachusetts, owned by Richard M. Sparrow. *Walt McCall Photo*

1932 Ahrens-Fox *Model "V"* 500-gpm Pumper, #7031, ex-Amherst, Ohio. *Walt McCall Photo*

1935 Ahrens-Fox 1,000-gpm *Model N-T-4* Piston Pumper #3415, Bare Cove Fire Museum, Hingham, Massachusetts. *Walt McCall Photo*

1936 Ahrens-Fox *Model YC Special Centrifugal* Pumper, #9007, ex-Greendale, Indiana, owned by Nathan Farloe. *Walt McCall Photo*

1937 Ahrens-Fox *Model I-C-U Quadruple Combination*, #9024, Everett, Washington. *Frank DeGruchy Photo*

1938 Ahrens-Fox *Model HT* 1,000-gpm Piston Pumper, #3446, ex-Hackensack, New Jersey, now owned by Maywood, New Jersey. *Dan Jasina Photo*

1939 Ahrens-Fox *Model VC* Centrifugal Pumper, #9079, Elfinwild-Glenshaw, Pennsylvania, owned by Mike Adams. *Dan Jasina Photo*

1939 Ahrens-Fox *Model BT* 1,250-gpm Pumper, #4024, with coupe style cab, Shelton, Connecticut. *John A. Calderone Photo*

1946 Ahrens-Fox Model HT 1,000-gpm Piston Pumper, #3459, First Hose Co., Hagerstown, Maryland. *Walt M. McCall Collection*

1947 Ahrens-Fox *Model HC Centrifugal* Pumper, #9102, 1,000-gpm, ex-River Rouge, Michigan, owned by Dennis Allenbaugh. *Walt McCall Photo*

1948 Ahrens-Fox *Model HT* 1,000-gpm Piston Pumper, #3476, ex-Lebanon, Pennsylvania, owned by Roger Engvall. *Walt McCall Photo*

1951 Ahrens-Fox *Model HT* 1,000-gpm Pumper with open semi-cab, #3486, Baldwin, Long Island, New York. *Martin Goffe Photo*

1955 Ahrens-Fox/GMC Commercial Pumper, ex-Melbourne, Kentucky, owned by Chip Lytle. *Walt McCall Photo*

1957 Ahrens-Fox *Model FCB* 750-gpm Cab-Forward Pumper by C.D. Beck, Ho-Ho-Kus, New Jersey. *Walt McCall*